Re-evaluation of the Indoor Resuspension Factor for the Screening Analysis of the Building Occupancy Scenario for NRC's License Termination Rule

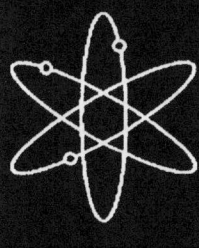

Draft Report for Comment

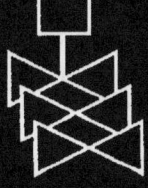

U.S. Nuclear Regulatory Commission
Office of Nuclear Material Safety and Safeguards
Washington, DC 20555-0001

AVAILABILITY OF REFERENCE MATERIALS
IN NRC PUBLICATIONS

NRC Reference Material

As of November 1999, you may electronically access NUREG-series publications and other NRC records at NRC's Public Electronic Reading Room at www.nrc.gov/NRC/ADAMS/index.html.
Publicly released records include, to name a few, NUREG-series publications; *Federal Register* notices; applicant, licensee, and vendor documents and correspondence; NRC correspondence and internal memoranda; bulletins and information notices; inspection and investigative reports; licensee event reports; and Commission papers and their attachments.

NRC publications in the NUREG series, NRC regulations, and *Title 10, Energy*, in the Code of *Federal Regulations* may also be purchased from one of these two sources.
1. The Superintendent of Documents
   U.S. Government Printing Office
   Mail Stop SSOP
   Washington, DC 20402–0001
   Internet: bookstore.gpo.gov
   Telephone: 202-512-1800
   Fax: 202-512-2250
2. The National Technical Information Service
   Springfield, VA 22161–0002
   www.ntis.gov
   1–800–553–6847 or, locally, 703–605–6000

A single copy of each NRC draft report for comment is available free, to the extent of supply, upon written request as follows:
Address: Office of the Chief Information Officer,
         Reproduction and Distribution
          Services Section
         U.S. Nuclear Regulatory Commission
         Washington, DC 20555-0001
E-mail: DISTRIBUTION@nrc.gov
Facsimile: 301–415–2289

Some publications in the NUREG series that are posted at NRC's Web site address www.nrc.gov/NRC/NUREGS/indexnum.html are updated periodically and may differ from the last printed version. Although references to material found on a Web site bear the date the material was accessed, the material available on the date cited may subsequently be removed from the site.

Non-NRC Reference Material

Documents available from public and special technical libraries include all open literature items, such as books, journal articles, and transactions, *Federal Register* notices, Federal and State legislation, and congressional reports. Such documents as theses, dissertations, foreign reports and translations, and non-NRC conference proceedings may be purchased from their sponsoring organization.

Copies of industry codes and standards used in a substantive manner in the NRC regulatory process are maintained at—
        The NRC Technical Library
        Two White Flint North
        11545 Rockville Pike
        Rockville, MD 20852–2738

These standards are available in the library for reference use by the public. Codes and standards are usually copyrighted and may be purchased from the originating organization or, if they are American National Standards, from—
        American National Standards Institute
        11 West 42nd Street
        New York, NY 10036–8002
        www.ansi.org
        212–642–4900

Re-evaluation of the Indoor Resuspension Factor for the Screening Analysis of the Building Occupancy Scenario for NRC's License Termination Rule

Draft Report for Comment

Manuscript Completed: April 20002
Date Published: June 2002

Prepared by
R. M. Abu-Eid, R. B. Codell
N. A. Eisenberg*, T. E. Harris
S. McGuire**

**Division of Waste Management
Office of Nuclear Material Safety and Safeguards
U.S. Nuclear Regulatory Commission
Washington, DC 20555-0001**

*N. A. Eisenberg
1208 Harding Lane
Silver Spring, MD 20905

**Incident Response Office
U.S. Nuclear Regulatory Commission
Washington, DC 20555-0001

COMMENTS ON DRAFT REPORT

Any interested party may submit comments on this report for consideration by the NRC staff. Comments may be accompanied by additional relevant information or supporting data. Please specify the report number NUREG-1720, draft, in your comments, and send them by August 30, 2002 to the following address:

Chief, Rules Review and Directives Branch
Division of Administrative Services
Office of Administration, Mail Stop T-6 D59
U.S. Nuclear Regulatory Commission
Washington, DC 20555-0001

You may also provide comments at the NRC Web site, http://www.nrc.gov. See the link under "Technical Reports in the NUREG Series" on the "Reference Library" page. Instructions for sending comments electronially are included with the document, NUREG-1720, at the web site.

For any questions about the material in this report, please contact:

Rateb (Boby) Abu-Eid
Mail Stop T-7 J8
U.S. Nuclear Regulatory Commission
Washington, DC 20555-0001
Phone: 301-415-5811
E-mail: BAE@nrc.gov

ABSTRACT

The purpose of this study was to re-evaluate the resuspension factor (RF) parameter used in the screening analysis for demonstration of compliance, using the building occupancy scenario, with the radiological criteria in the license termination rule in 10 CFR 20, Subpart E. The RF is a highly sensitive parameter impacting the inhalation dose calculation. An RF parameter value of 1.42×10^{-4} m^{-1} was established for screening analysis (Beyeler et al, 1999). Assuming a 10 percent fraction of loose (removable) contamination, NRC staff selected a default RF value of 1.42×10^{-5} m^{-1} for use in the inhalation dose calculation. Based on this RF value, and using the DandD code, the derived default concentration or surface activity screening limits for most radionuclides, particularly the alpha-emitters, were at background levels or far below the corresponding detection limits. In this study, NRC staff analyzed further literature data considering more realistic assumptions of the average member of the critical group in the building occupancy scenario and accounting for more recent actual RF field data collected for two facilities undergoing decommissioning. Based on the current analysis and re-evaluation, staff recommends using an RF value of 10^{-6} m^{-1} in the screening analysis of the inhalation dose calculation for the building occupancy scenario. The staff believes that the newly proposed RF default value is more realistic than the current value in DandD code, and sufficiently conservative for screening analysis.

CONTENTS

CONTENTS (cont'd)

LIST OF FIGURES

LIST OF TABLES

EXECUTIVE SUMMARY

This study was conducted to re-evaluate and establish a more realistic and representative resuspension factor (RF) for use in screening dose analysis. Based on a study conducted by Sandia National Laboratory, (SNL) (Beyeler et al, 1999), NRC staff adopted a default RF value of 1.42×10^{-5} m^{-1} for use in DandD screening code to derive default concentrations or surface activity screening limits (NUREG-1727, 2000). Due to the highly conservative value of the RF, the derived surface activity levels for most alpha-emitting radionuclides were unrealistically low at or near background levels or below the corresponding detection limits. For example, the screening concentrations equivalent to 0.25 mSv/y (25 mrem/y) for Th-232, U-238, and Am-241 were derived at 0.12, 1.68, and 0.45 Bq/100 cm^2 (7.3, 101, and 27 dpm/100 cm^2) respectively.

In this study, the staff evaluated the main factors affecting the RF value such as the driving forces, the removal mechanisms, the characteristics of surface activity (e.g., bound or loose), and the particle size. Staff assessed these factors considering the building occupancy scenario as defined in NUREG/CR-5512, Volume 1 (NRC, 1992); NUREG-1496, (NRC, 1997); and NUREG-1549, (NRC, 1998). In addition, staff assessed current tests used to determine removable (loose) fraction using the "wipe" or "smear" tests. Further, the staff critically evaluated the basis for deriving the indoor RF in NUREG/CR-5512, Volume. 1, and SNL approach (Beyeler, et al, 1999) for development of the RF default value in DandD code Versions 1.0 and 2.1. The study also evaluated published RF data applicable to the building occupancy scenario in consideration of the representativeness of such data to decommissioning sites conditions particularly regarding the driving forces, ventilation, and surface activity adhesion conditions. More importantly, staff analyzed and evaluated measurements of surface activity and airborne activity concentration for facilities undergoing decommissioning.

Using published literature data and extensive field measurements at two decommissioning facilities, the staff used statistical analysis to evaluate time variation of airborne concentration, conducted tests of independence of data from different locations, assessed partitioning of data, and evaluated tolerance limits. As a result of the staff's re-valuation of the RF data, an improved basis to estimate indoor RF has been established. Finally, the staff conducted statistical analysis of RF mean values for five sites (e.g., five data points) deemed applicable to the building occupancy scenario as well as to decommissioning site conditions. The staff believes that the available data and information on these sites are not perfect, but they provide the best insight available at the present time to estimate the probability density function (PDF) for the RF. Overall, the authors of this report believe these data provide an overestimate of the distribution of RF likely to exist at decommissioned facilities. We deemed it appropriate to base the PDF for RF on the 5 data points representing the site means, adjusted for worker occupancy, because: (1) workers may move around a facility and be exposed to a variety of air concentrations; and (2) the regulation is written to protect the average member of the critical group. We fitted the five site data to a normal and a lognormal distribution. Since there were only five data points, we felt that it was appropriate to use the "maximum likelihood" approach (Benjamin and Cornell, 1970) to estimate the distribution rather than a statistical (i.e., "unbiased") approach. The difference between the two approaches is that the estimated standard deviation in the maximum likelihood approach is smaller by the ratio $\sqrt{(N-1)/N}$. This smaller standard deviation will lead to a slightly smaller value for the 90th percentile of the distribution, which is used as the suggested regulatory criterion for RF. The parameters of the normal and lognormal distributions for the maximum likelihood fits are given below:

Parameters for Normal and Lognormal "Maximum Likelihood" Models of RF Data

Statistical Model	Sample Mean	Sample Standard Deviation	90th Percentile RF
Normal Fit to 5 site mean RF's	4.74×10^{-7} m^{-1}	3.11×10^{-7} m^{-1}	8.7×10^{-7} m^{-1}
Lognormal Fit to 5 site mean RF's	$\log_{10} = -6.433$	$\log_{10} = 0.3247$	9.6×10^{-7} m^{-1}

Although both the normal and lognormal distributions are reasonable fits to the data, the normal distribution has the disadvantage of allowing negative values of RF, which is not physically possible. In addition, the lognormal fit is more conservative choice at the 90th percentile RF.

This study resulted in a recommendation of using an RF value of 10^{-6} m^{-1} for screening dose analysis as an alternate to the current default value 1.42×10^{-5} m^{-1} used in the NRC's DandD code Version 2.1. This recommendation was based on rounding the nominal 90th percentile of the PDF RF value (e.g., 9.6×10^{-7} m^{-1}) using a lognormal fit.

FOREWORD

This report is a product of the staff's continuing efforts to establish more realistic and representative default values for use in screening performance assessment or dose analysis approaches. The current study was specifically conducted to re-evaluate the default screening value of the resuspension factor (RF) parameter used in decommissioning screening analysis. The RF is a highly sensitive physical parameter that impacts the calculated inhalation dose and subsequently the derived dose limit used for demonstration of compliance with NRC's license termination rule for decommissioning (10CFR20, Subpart E). The RF parameter is difficult to determine in a realistic and reliable fashion because it requires extensive and costly measurements over a long time period. Therefore, the staff attempted to critically evaluate published RF data, deemed applicable to the building occupancy scenario, and use more recent empirical field data collected over 1-3 years at two facilities owned by Westinghouse Electric Company and BWX Technologies, Inc. Based on the staff's current analysis and evaluation, the RF default screening value for the building occupancy scenario may be reduced by an order of magnitude.

This draft NUREG report is not a substitute for NRC regulations and compliance with it is not required. The approaches and/or methods presented in this NUREG are provided for information only. The report is intended to solicit comments and feedback on staff analysis and approaches, and to explore availability of more recent field or experimental indoor RF data that may be used to optimize the current default RF value. Publication of this report does not necessarily constitute NRC approval or agreement with the information contained herein. Use of product or trade names is for identification purpose only and does not constitute endorsement by the NRC.

Sandra L. Wastler, Acting Chief
Environmental and Performance Assessment
 Branch
Division of Waste Management
Office of Nuclear Material Safety and Safeguards

ACKNOWLEDGMENTS

The authors wish to express their thanks to several people who helped in the development of this report. Joseph Nardi of Westinghouse Electric Company and David Spangler of BWX Technologies, Inc., provided significant field monitoring data on resuspension factor, in response to discussions held at the U.S. Nuclear Regulatory Commission's public workshops during 1999 -2000. Lee Abramson, Office of Nuclear Regulatory Research, NRC, provided invaluable help in the statistical interpretation of the sparse data on resuspension. Duane Schmidt, Office of Nuclear Material Safety and Safeguards, NRC, and James Weldy, Center for Nuclear Waste Regulatory Analyses, conducted reviews and provided valuable comments that helped improve early draft of this report.

1.0 INTRODUCTION

The U.S. Nuclear Regulatory Commission's (NRC's) "Final Rule on Radiological Criteria for License Termination" (NRC, 1997) requires that, in order to terminate a license, the dose to the average member of the critical group from residual radioactivity distinguishable from background must be no greater than 0.25 mSv per year (25 mrem/yr). In addition, this rule requires that the residual radioactivity has been reduced to levels that are as low as is reasonably achievable (ALARA).

For residual radioactivity on building surfaces, the concentration that would result in a dose of 0.25 mSv per year (25 mrem/yr) to the average member of the critical group may be calculated using the screening building occupancy scenario described in NUREG/CR-5512, Volume 1 (Kennedy and Strenge, 1992, Section 3.2). The building occupancy scenario for screening assumes that light industrial activities will take place in the decommissioned building (NRC, 1998 and NRC, 2001). The building occupancy scenario assumes three pathways by which a future occupant of the building can be exposed to radiation. These pathways include: direct external radiation; inhalation of residual radioactivity resuspended from surfaces, because of activities of occupants; and ingestion of the residual radioactivity wiped off the surfaces and subsequently ingested by occupants. The NRC is currently using the computer code DandD, version 2.1, to perform screening analyses (NRC, 2001).

In evaluating the generic screening values, using DandD for the building occupancy scenario, it was apparent that the values for alpha-emitters were very low, in many cases below detection levels. Consistent with Commission direction for NRC staff to evaluate excessive conservatism in the DandD code, we evaluated the causes for these very low values and whether there was excessive conservatism. Based on our evaluation, we determined that the indoor resuspension factor (RF) was one aspect of the methodology where excessive conservatism may have contributed to the very low screening values.

For many nuclides, in particular the important alpha-emitters such as uranium and thorium, the inhalation pathway is typically the predominant exposure pathway. The RF is the most sensitive parameter affecting the inhalation dose. In the inhalation pathway model incorporated into the DandD, the RF is the only factor treated as a random variable during selection of the default parameter values.

In Section 2, this report discusses how the RF is used in the inhalation dose calculation and the factors affecting the RF. Section 3 discusses how the default values were selected for NUREG/CR-5512 and the DandD code. Section 4 provides a summary and evaluation of studies of the RF. Section 5 discusses the development of an alternate RF for decommissioning cases. Section 6 presents conclusions and recommendations of this report regarding selection of a screening value for RF.

2.0   INHALATION DOSE CALCULATIONS AND FACTORS AFFECTING THE INDOOR RF

2.1   The NUREG/CR-5512 and the DandD Inhalation Dose Model

The DandD computer code uses the same equation as presented in NUREG/CR-5512, Volume 1 (Kennedy and Strenge, 1992, Volume 1) to calculate the inhalation dose for the building occupancy scenario. That is, the dose from inhalation can be calculated by:

$$D_{inh} = DCF_{inh} \times B \times t \times RF \times C_{surf} \qquad (1)$$

D_{inh} is the committed effective dose equivalent rate from inhalation, mSv/y (mrem/yr),
DCF_{inh} is the dose conversion factor for the radionuclide inhaled, mSv/Bq (mrem/pCi),
B is the breathing rate, m³/hr (ft³/hr),
t is the annual occupancy time, hr/yr,
RF is the indoor RF relating airborne concentration to surface concentration, m⁻¹ (ft⁻¹), and
C_{surf} is the surface concentration, becquerel per meter square, Bq/m² (pCi/m²)

The DCF is a fixed value from Federal Guidance Report No. 11 (U.S. Environmental Protection Agency, 1988). The breathing rate and annual occupancy time are metabolic and behavioral parameters that are fixed based on assumptions made in developing the critical group and default scenario. The surface concentration is a measured site specific parameter. The RF value is a variable dependent on several factors. The RF is considered to be a random variable whose distribution represents the range of conditions (both physical conditions of the contamination and the behavior conditions leading to resuspension) that might be found at sites that have undergone decontamination. Unlike other dose models in DandD, the indoor inhalation-dose model for building occupancy scenario generally allows only one random variable, RF, that affects dose.

2.2 Factors Affecting the RF

The RF is the ratio of the airborne concentration of contamination to the surface concentration of contamination. The RF is affected by a number of physical factors that include: type of disturbance, intensity of disturbance, time since deposition, nature of the surface, particle size distribution, climatic conditions, type of deposition, chemical properties of the contaminant, surface chemistry, and building geometry and physical characteristics. A general discussion of these factors is provided in NUREG/CR-5512, Volume 3 (Beyeler, et al., 1999).

In the simplest terms, the RF is determined considering the nature of contamination on the surface (e.g., how tightly bound to the surface it is), and balancing the driving forces that cause the material on the surface to become airborne, and the mechanisms that remove the material from the air. Particle-size effects also play an important role in the airborne concentration of contaminates, and thus the RF. In assessing these factors, one must consider the circumstances under which the RF will apply (i.e., activities, physical conditions, and structures associated with the building occupancy scenario). Clearly, the concept of RF applies to particulate solids and does not apply to gases.

2.2.1 Driving Forces

The primary driving force that will resuspend particles in the building-occupancy scenario can be expected to be mechanical forces associated with rubbing and abrasion of surfaces. These forces are typically caused by the activities or movements of the occupants like walking and moving carts (Corn and Stein, 1967; Morton, 1999). In buildings, air currents caused by normal room ventilation or by vibrations are not expected to be a major cause of resuspension of particles (Walker et al., 1967; Hinds, 1982). Moreover, RFs determined from mechanical disturbance can be one order of magnitude higher than RFs determined with air currents only (Beyeler et al., 1999). Higher RFs were measured when driving forces were increased and when the surface contamination was loose or easily removable. Several studies of RF, including Fish, et al. (1967), observed a power-law relationship between air velocity and RF. Fish, et al. (1967) also reported a difference in the RF of greater than an order of magnitude due to the type of driving forces. Jones and Pond (1967) also reported variations in the RF from different walking speeds. Therefore, it is important to assess the types and intensity of the applied driving forces to evaluate the corresponding RF measurements, to determine if they are reasonably representative of the building-occupancy scenario.

For the building-occupancy scenario, driving forces (worker activities/movements) should simulate normal workplace activities that would occur over an entire average working year. This can best be accomplished if measurements are made while normal activities are being conducted or if actual worker activities/movements are observed and reproduced faithfully. The RF measurements of activities done for only brief periods should not be assumed to be representative of RF measurements made over long periods of time.

2.2.2 Removal Mechanisms

In assessing studies that are representative of the building occupancy scenario, consideration must be given to room ventilation. Although ventilation does not cause significant resuspension, it will cause removal of already suspended particles by two mechanisms. The first removal mechanism is by outflow of air from the room. The second removal mechanism is turbulent inertial impaction caused by the change in direction of air streams as the air goes around the obstacles in the room or in the ventilation system. These removal mechanisms are important because they will reduce the airborne concentration and thus the RF.

For the building occupancy scenario, it can be assumed that the ventilation for a light industrial facility would meet national codes and standards (e.g., ASHRAE, 1989) as well as State and local requirements. Thus, to be representative of the building-occupancy scenario, measurements should be conducted with ventilation similar to those found at light industrial-use facilities. Measurements taken with no room ventilation will likely overestimate the RF because the primary mechanisms for removal of airborne particulates were not present. Similarly, measurements taken with excessive room ventilation are likely to underestimate the RF.

3

2.2.3 Characteristics of the Surface Activity

The characteristics of how the surface activity is bound to the surface will have a major effect on the RF. For particles to become resuspended, the bond between the particles and the surface (e.g., floor) must be broken by the driving forces (i.e., mechanical or air forces). Particles that are tightly bound to the surface are not easily resuspended whereas particles that are loosely bound, like freshly deposited material, will be more easily resuspended.

The adhesion of particles has been studied extensively, and although it is a very complex process, the general principles are well understood. Hinds (1982) related the main surface adhesion forces to either van der Waals force, electrostatic force, and/or surface tension forces of adsorbed liquid films. These forces are affected by the material type, shape, and size of the particles. In addition, the material roughness, the relative humidity, temperature, duration of contact, and initial contact velocity are important factors affecting surface adhesion. The most important adhesion forces are the London-van der Waals forces, the long range attractive forces that exist between molecules. In general, adhesive forces are inversely proportional to the diameter of particles "d" while removal forces are proportional to d^3 for vibration and centrifugal force or to d^2 for air currents. This suggests that as the size of particles decreases, it becomes increasingly difficult to remove them from the surfaces. For example, the adhesive forces on particles of less than 10 μm are much greater than other forces that such particle experience.

All small particles generally adhere to and are bound to the surface, and no particles are really "loose." Therefore, particles are removed from surfaces almost entirely by applying a mechanical force to the particle sufficient to break the adhesive bond. Particles that are loosely bound to the surface will be easily removed and resuspended. Particles that are tightly bound require greater mechanical force to break the bonds and become resuspended. If the bond is not broken, then the particle will not become resuspended. Therefore, the nature of the contamination on the surface will have a important effect on the RF. For the same amount of total surface activity, surfaces with a large portion of loosely bound particles would be expected to have a larger RF, and surfaces where almost all the particles are tightly bound would be expected to have a smaller RF. The amount of loosely bound particles could change as the surface degrades over time with application of mechanical forces.

NUREG/CR-5512, Volume 3 (Beyeler, 1999) reported that "several studies model variations of resuspension factor with time, including Kathren (1968), Langham (1969), NRC (1975), IAEA (1982, 1986), Garland (1982), and Nair, et al., (1977)". All of these models produced decrease in RF with time, reflecting the experimentally observed decrease in contaminant air concentration with time over contaminated areas. This trend also explained that contaminants become more fixed with time and the contaminated source on surfaces becomes more depleted with time.

Consideration of the representativeness of the surface activity is important in selecting measurements that are applicable to decommissioned facilities. We consider that good housekeeping practices will be used in normal decommissioning as a minimum to meet the ALARA requirements[1] in 10 CFR 20.1402. It is assumed that surfaces will be cleaned or

---

[1]ALARA requirements are further discussed in DG-4006 and in Section 7 of the Standard Review Plan.

washed during decommissioning. This will remove most of the loosely bound and some of the more tightly bound particles. Following the above discussion, surfaces that have been cleaned would be expected to have a smaller RF than surfaces that have not been cleaned, given the same levels of surface contamination.

2.2.4 Particle Size Effects

Particles are often classified by their activity mean aerodynamic diameter. This value provides information about the particles aerodynamic behavior and how the particles will deposit in the respiratory system. The particle diameter is typically expressed as the mean diameter. It is common practice to consider respirable particles (i.e., particles able to reach the pulmonary region of the lung) as having a mean diameter of 10 µm or less. It is therefore most important to evaluate the activity of particles that are respirable. Larger particles typically do not reach the pulmonary region of the lung and may be exhaled without contributing to inhalation dose, ingested, or otherwise absorbed, leading to doses other than to the lung (Cember, 1996).

Fish, et al., (1967) reports a strong correlation of RF with particle diameter. As discussed in NUREG/CR-5512, Volume 3, resuspension is greatest for smaller diameter particles. The RF decreases with particle diameters in the range of 1 to 5 µm. As discussed above, this also corresponds to the particle- diameter range that provides the most significant dose. In addition, the distribution of particle sizes may change over time as mechanical forces are applied.

Although larger-diameter particles may be resuspended, gravitational settling removes them from the air more rapidly than smaller particles. Nevertheless, larger particles can be important because they can be measured as "removable" by a wipe test, leading to the conclusion that a higher fraction of resuspendable particles may be present than can actually contribute to dose. In the context of this report, which is the estimation of RF's representative of decommissioned buildings, significant removable activity as larger particles may cause the RF to be under-estimated. Since RF is a ratio, the numerator is set equal to the measured air concentration, whereas the denominator is set equal to the measured surface activity.

Information about the mean airborne particle size is usually not provided in studies presenting resuspension data. However where information is provided on particle-size distributions (e.g., on the air samplers or surface samplers), it is important to weigh the effect on the estimated RF.

2.3 Using the Wipe Tests to Assess Removable Fraction

Particles on surfaces are sometimes described as being of two types: (1) "fixed," "bound" or "non-removable" particles; and (2) "loose," "unbound," or "removable" particles. The "smear" or "wipe" measurement is often taken to be a measurement of the particles that are "loose." In reality, this distinction is not exact, but it can be useful with proper understanding of the underlying process.

The wipe test provides information about the fraction of the particles that projects high enough above the surface to be subjected to the mechanical forces of the wipe. Basically almost all particles physically touched by the surface of the wipe will have their bonds broken because the force used for the rubbing will be far greater than the particle bond strength. A wipe will break the bonds of many of the particles that are on the microscopic peaks on the surface profile, but will affect few particles in the valleys and depressions in the surface. After the bonds are broken, the particles can then either re-attach themselves to the surface at another location, re-attach to the wipe material, or become airborne. This latter event requires that the particles have sufficient kinetic energy to overcome the van der Waals and electrostatic forces and that they have a free pathway for escape. Hence, a wipe measurement usually includes more than "loose" activity. Considering this analogy, a wipe test may not adequately represent the fraction of particles that would be resuspended by walking.

3.0 PREVIOUS DETERMINATIONS OF THE RF

3.1 Basis for Deriving the Indoor RF in NUREG/CR-5512, Volume 1

NUREG/CR-5512 Volume 1 (Kennedy and Strenge, 1992), recommended a specific value for each of the parameters in equation 1. The recommended value for the indoor RF was 10^{-6} m^{-1}. However, there was no detailed explanation of how the value was determined. William Kennedy, the principal author of Volume 1, revealed (Kennedy, 1999) indicated that the authors relied, in part, on Brodsky (Brodsky, 1980), who concluded that, although vigorous disturbances could produce RFs higher than 10^{-6} m^{-1}, normal activities averaged over long periods of time would have RFS of less than 10^{-6} m^{-1}. The Volume 1 authors also relied on their own experience and background knowledge in leading them to conclude that 10^{-6} m^{-1} is an upper bounding limit under ordinary conditions that would be expected at a decommissioned facility.

3.2 Development of the RF in DandD Code Version 1.0

Unlike the deterministic value used in NUREG/CR-5512, Volume 1, the RF was treated probabilistically in establishing the default parameters for the DandD code, version 1.0. The approach used to develop the default RF parameter in DandD code is documented in Volume 3 (Beyeler, et al., 1999). A distribution describing the variability of the RF (i.e., a probability density function (PDF)) was established.

As described in NUREG/CR-5512, Volume 3, Sandia National Laboratories (SNL) reviewed a number of studies published between 1964 and 1997, and determined that only a small number of studies provided numerical results pertinent to indoor resuspension for the building-occupancy scenario. Reported RF values from all these studies ranged from 2×10^{-8} to 4×10^{-2} m^{-1}. Some of these studies were deemed inapplicable, for the following reasons: (1) the study did not provide results that could be converted to an RF; (2) the study conditions included sources of airborne contamination other than resuspension; (3) the contaminated surface in the study (e.g., clothing) was not representative of building surfaces; or (4) the mechanical stresses on the contaminated surfaces were not representative of the conditions in the building occupancy scenario.

NUREG/CR-5512, Volume 3, concluded that two RF studies (Jones and Pond, 1967; Fish, et al., 1967) were applicable. For both of these studies, the surface contamination was freshly deposited (by the researchers). Based on the assumption that, in these studies, essentially all the contamination was removable, SNL expressed the RF for a decommissioned facility as the product of the RF for loose, or removable, contamination and the fraction of the total contamination that was removable.

The data for RF were categorized by similarity of the nature (air flow and mechanical disturbance) and intensity (low or high air flow, absence or presence of mechanical disturbance) of the surface disturbance. Three categories were used: (A) low air flow and no mechanical disturbance; (B) low air flow with mechanical disturbance; and (C) high air flow with mechanical disturbance. Data from the two studies were grouped into these categories, and ranges (minimum and maximum) of the RF were described for each category. Values from Category "C" were adjusted to an effective value to account for the source depletion that would occur at a high RF and high ventilation rate (high air flow).

SNL acknowledged that the RF values from these studies represented pessimistic estimates, and the spread of data would likely overestimate the uncertainty in an annual average RF. However, SNL pointed out that with such a limited number of studies, the existing data were not likely to describe the full range of potential RF values. SNL concluded that these two effects tend to counteract each other, and that correction for the effects was not reasonable with a limited pool of data. SNL adopted the pessimistic values as estimates of the annual average RF values.

To combine results from the categories to form a PDF, NUREG/CR-5512,Volume 3, estimated the fraction of light industrial structures that would fit the conditions for the categories. This weighting was determined to be 0 percent for Category A (because the lack of mechanical disturbance was seen as inconsistent with light industrial use); 90.2 percent for Category B, and 9.8 percent for Category C. Loguniform distributions were assumed for the RF for categories B (with minimum 9.1×10^{-6} m^{-1} and maximum 1.9×10^{-4} m^{-1}) and C (with minimum 7.1×10^{-6} m^{-1} and maximum 1.4×10^{-4} m^{-1}). Based on these distributions, and the category weighting, the resultant PDF for the RF for removable contamination was developed as shown in Figure 1. This resultant PDF ranges from 9.1×10^{-6} m^{-1} to 1.9×10^{-4} m^{-1}, with median value 5.0×10^{-4} m^{-1} and default value for the DandD code (90th percentile) of 1.42×10^{-4} m^{-1}.

Finally, to calculate the RF for decommissioned sites, the fraction of total contamination that is loose (removable) must be addressed. In this respect, NRC staff has assumed that a reasonable value for screening purposes is 0.1. This removable fraction value has been used to develop a DandD default parameter value of 1.42×10^{-5} m^{-1} applicable for all surface contamination types (e.g., removable and non-removable) of decommissioned sites.

4.0   REVIEW AND EVALUATION OF MEASURED DATA FOR THE INDOOR RF

This section reviews measurement studies of the indoor RF. If one considers all the possibilities, the RF will have a value ranging from zero (when there is no driving force to disturb the surface) to very large values (if there is a vigorous mechanical force such as scraping or grinding applied on the surface). However, if we consider only those measurements representative of long-term activities that represent the building-occupancy scenario, then the range of the indoor RF distribution may be greatly narrowed. Furthermore, although some vigorous activities may result in peaks of air concentration, what is of interest is the annual dose which is related to the average conditions for a year. In selecting experiments to determine the RF for the building- occupancy scenario, it is necessary to use measurements that are representative of the building- occupancy scenario. This means that the driving force, the ventilation (removal mechanism), particle size, and the degree to which the material is bound to the surface should all be appropriate and compatible with the scenario.

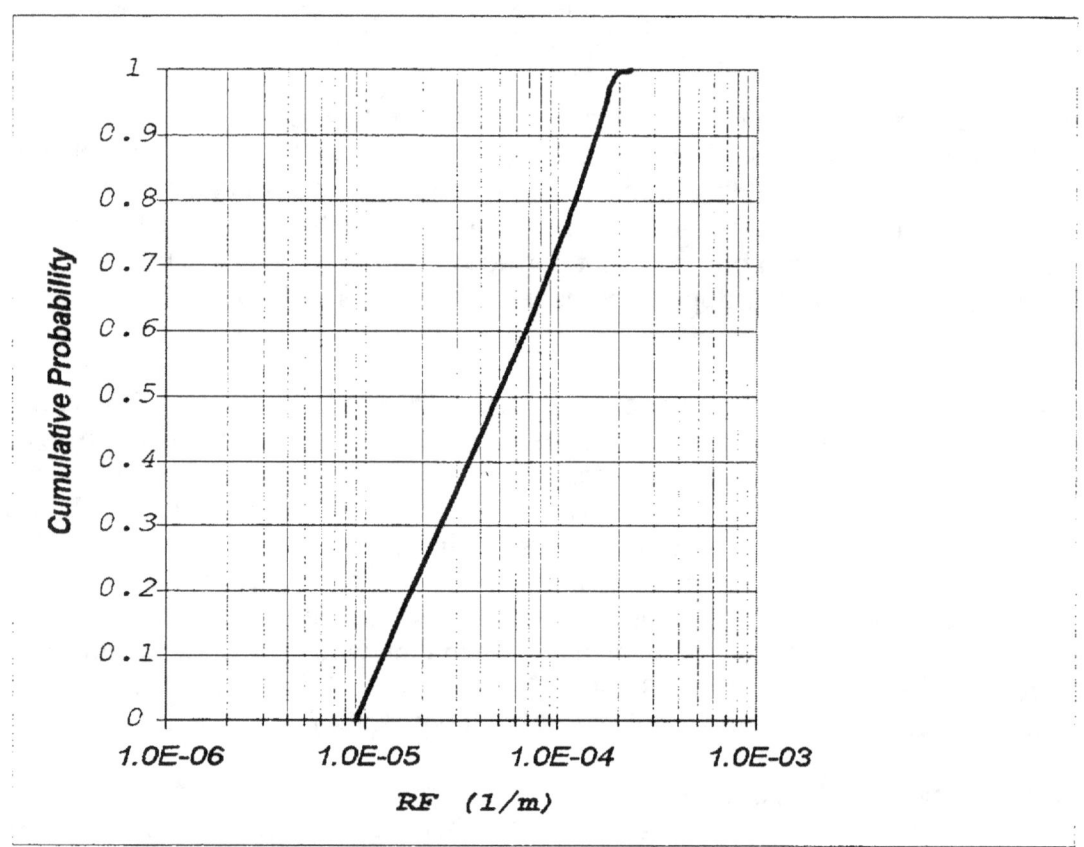

Figure 1. **Cumulative Probability Function for RF Developed for DandD Code by SNL (Beyeler, et al. (1999))**

1   We present below a brief description and our conclusions regarding applicability and compatibility
2   of each study to decommissioned facilities. We also address factors that might tend to
3   overestimate or underestimate the RF value applicable to a building-occupancy scenario. In this
4   regard, there are three major criteria that need to be considered when assessing
5   representativeness of the RF data for decommissioned sites:
6
7   a)   The RF data should have been derived using a driving force representative, to the extent
8   practicable, of the decommissioning facilities (e.g., similar to activities of the light-industry
9   scenario which is the critical group for the building-occupancy scenario);

| | | b) | The RF data should have been generated for facilities with ventilation conditions as similar as practicable to the light industry scenario described above. Thus, data generated under forced or abnormal ventilation conditions (e.g., directing fans or hair dryers towards loose contamination on the floor) or under no ventilation or air flow were rejected; and |

1
2
3
4
5
b) The RF data should have been generated for facilities with ventilation conditions as similar as practicable to the light industry scenario described above. Thus, data generated under forced or abnormal ventilation conditions (e.g., directing fans or hair dryers towards loose contamination on the floor) or under no ventilation or air flow were rejected; and

6
7
8
9
c) The surface activity should be assumed to adhere to the surface to some extent or assumed to occur on surfaces that went through cleaning or washing processes. These assumptions are used because ALARA requires cleaning or washing of contaminated surfaces for facilities undergoing decommissioning.

10
11
12
13
14
15
16
Table 1 summarizes the representativeness for different studies of the driving forces, the room ventilation conditions, and surface activity adhesion as applicable to the building-occupancy scenario. The representativeness of the surface activity to decommissioned sites (i.e., cleaned with a small percentage of loosely bound contamination) is presented. In Table 1, a "+" is used to indicate conditions that are representative of decommissioned facilities, and a "-" is used to indicate conditions that are not representative. An "0" is used to indicate that either the conditions are mixed or not sufficiently documented to assess.

17

Table 1: Summary of Representativeness of RF Data

Study	Representativeness of:		
	Driving Forces	Room Ventilation	Surface Activity Adhesion
Breslin, 1966	+	+	0
Eisenbud, 1954	+	+	0
Fish, 1967	0	-	-
Jones, 1967	0	+	-
Ikezawa, 1980	+	+	-
Nardi, 1999	+	+	+
Ruhter, 1988	+	0	+
Spangler, 1998	+	+	+

18 · 19 · 20 · 21 · 22 · 23 · 24 · 25 · 26

27
28
29
+ conditions representative of decommissioned facility.
- conditions not representative of decommissioned facility.
0 conditions are mixed or not sufficiently documented to assess.

30
31
Another factor that could be considered in evaluating the studies is the improvements in measurement instruments and calibration techniques over time. Calibrations of alpha activity

measurements of surface activity, conducted during the early studies (e.g., 1954 - 1967), under-estimate the total surface activity by about a factor of two (Abelquist, et al., 1998). Thus, it is likely that the RF value is significantly over-estimated for the early studies. In addition, modern survey instruments are more sensitive and will more accurately measure activity. However, we will not use this factor of two to adjust downward any of the RF estimates from the early studies.

4.1   Breslin, 1966, Data

4.1.1 Description

RF data were collected at an operating uranium processing plant over a weekend while operations were not being conducted. However, the surrogate workers attempted to duplicate normal working activities and movements that they had observed during operation. These data represent three different operational areas designated as: the "Assistant Press Operator" area, the "Rod Puller" work area, and the "Uranium Extrusion" area. Operational activities at the uranium processing plant introduced a significant amount of airborne activity. For each area, there were four measurements taken relative to operations: (a) no operational impacts (i.e., airborne contamination introduced by operations had settled out of the air); (b) post-operation transient conditions (i.e., airborne contamination introduced by operations had not completely settled out of the air); (c) initial operating transient conditions (i.e., operations had begun to introduce airborne contamination but had not yet reached equilibrium); and (d) operational conditions (i.e., equilibrium of airborne contamination introduced from operations had been reached). Two data points, representing moving and work practices of two different workers, were reported under each of these conditions, for each of the three facilities for a total of 24 data points. In addition, the study reported one data point at each of the four conditions for each work area for a total of 12 data points. The averages of surface activities for the three facilities were measured at 3.0×10^4 Bq/m^2 (1.8×10^6 dpm/m^2)for the Assistant Press Operator facility and 8.3×10^4 Bq/m^2 (5×10^6 dpm/m^2) for each of the other two areas. A summary of the data is provided in Appendix A.

4.1.2 Evaluation

In assessing the RF data relative to decommissioning sites, we considered that data under Condition (a) were representative of decommissioned sites. Some of the data listed under Condition (b) may correspond to decommissioned conditions. The remaining data show significant interferences arising from airborne contamination introduced by operations and were not considered to be representative of decommissioned facilities. The average RF values corresponding to Condition (a), for each of the three areas and for each of the three air samplers for a total of nine values, were used in the evaluation of RF.

We note that some of the data were collected by lapel (Breathing Zone) samplers worn by two different workers, and some of the data were collected using a two-stage sampling instruments at a fixed location within the facility. There was clearly a difference between the lapel samplers and the fixed samplers, with the former being significantly higher (an average of approximately 28 percent for the data used).

Several factors will cause the data from this study to potentially overestimate the RF at decommissioning sites. First, workers' activities and movements during the experiment were

10

1	conducted in an exaggerated active manner to maximize resuspension to determine an upper
2	bound on resuspension. Second, more loose residual radioactivity was present than would be
3	anticipated at a decommissioned facility, making the observed resuspension larger than the
4	resuspension at a decommissioned facility, as demonstrated by the observation of the fall-off of
5	airborne concentrations with time (this is discussed further in Section 5.1.1). Therefore, the data
6	should be used with the understanding that the RFS are overestimated by some factor that
7	cannot be precisely quantified.

8 ## 4.2   Eisenbud, 1954, Data

9 ### 4.2.1   Description

10	Airborne radioactivity concentrations during plant operations were compared with different
11	surface radioactivity concentrations at several operating uranium and radium processing
12	facilities. The purpose of the study was to estimate the importance of surface activity for causing
13	airborne activity. The Eisenbud, et al., (1954) study concluded that airborne concentration is
14	attributable mainly to operational activities rather than resuspension from surface activity.

15	Several areas within the uranium and radium processing facilities were studied. As with the
16	Breslin (1966) study, operations at these facilities introduced a significant amount of airborne
17	contamination. In addition, most of the areas had very low surface activity. Therefore, the
18	airborne contamination is attributed to operational effects. However, one area (Plant J) did have
19	a high surface activity and low operational airborne contamination.

20 ### 4.2.2   Evaluation

21	We consider that data from Plant J are marginally representative of decommissioned facilities.
22	The remaining data show significant interferences arising from airborne contamination caused by
23	operations and were not considered to be representative of decommissioned facilities. For plant
24	J, three RF values were reported: 0.1 x; 0.32 x; and 0.50 x 10^{-6} m^{-1}.

25	The assumption that all airborne activity at this site is derived from resuspension will tend to
26	overestimate the RF because particulate activity is largely influenced by ongoing operations.
27	Also, there had not been much cleaning of surfaces so that there is likely to be more
28	resuspension than would occur from a cleaned surface. However, this study suggests that the
29	average value of 0.3 x 10^{-6} m^{-1} could perhaps be near the high end of the RF distribution for the
30	building- occupancy scenario.

31 ## 4.3   Fish, 1967, Data

32 ### 4.3.1   Description

33	RF values were developed from experimental conditions. Zinc sulfide (ZnS$_2$) and cupric oxide
34	(CuO) particles were freshly dispersed in a test room with painted drywall walls and asphalt tile
35	floors. There were four sets of measurements:

36	1.	Ten minutes of vigorous activity, including sweeping with no exhaust or fans. The
37		estimated RF for was 190 x 10^{-6} m^{-1}.

2. Twenty minutes of vigorous walking. The estimated RF was 39×10^{-6} m^{-1}.

3. Forty minutes of light work activity. The estimated RF was 9.4×10^{-6} m^{-1}.

4. Ninety minutes of some light sweeping and some other light activity with no exhaust ventilation, but with fans for circulation. The estimated RF was 710×10^{-6} m^{-1}.

We consider that the driving forces for measurements 1 and 4 are not representative of a light industrial facility. In addition, the fourth measurement appears to be an outlier with respect to the other measurements reported in the study and with respect to the other studies described in this report.

4.3.2 Evaluation

We do not consider this study to be representative of decommissioned sites for the following reasons: 1) There was no ventilation to reduce the airborne concentrations; 2) The surfaces had not been washed or otherwise treated to remove the easily removable particles; 3) The densities of ZnS_2 and CuO are lower than most radionuclides of interest, particularly uranium and transuranics; and 4) Driving forces and measurement techniques were not always representative (for example, certain data were obtained with air samplers located near the floor and extreme air circulation was produced by fans aimed at the floor). These factors will cause the measured RF to be overestimated, for the purposes of decommissioned facilities. However, the magnitude of the difference cannot be determined.

4.4   Ikezawa, 1980, Data

4.4.1   Description

The Ikezawa data were generated to assess the procedure of decontamination and cleanup levels immediately after an accidental break of negative pressure in a plutonium (Pu) hot-cell glove box. Airborne concentrations were measured by personal air samplers on two workers who engaged in cleanup work. The measurements were conducted before any cleanup or remedial actions. The released aerosol particulates were easily suspended due to this instantaneous and fresh release of contaminants.

This study reported a mean RF of 180×10^{-6} m^{-1} for decontamination activities of floors and walls. A mean RF value of 2.3×10^{-6} m^{-1} was reported when no work was being performed. A range of RF values (4 to 20×10^{-6} m^{-1}) was also reported for decontamination activities of a hot cell.

4.4.2   Evaluation

This study is not considered to be representative of decommissioned facilities. The surface activity was freshly deposited powder, which is not representative of cleaned decommissioned facilities. As discussed in Section 2.2.4, the large amount of readily removable activity will likely cause the RF to be greatly overestimated.

4.5   Jones, 1967, Data

4.5.1   Description

Jones studied the resuspension of plutonium oxide and plutonium nitrate from floors. These materials were deposited on the floors as a water suspension that was subsequently left to dry.

The floor materials used in the experiment included: wax paper, PVC sheet, waxed linoleum, and unwaxed linoleum. The investigators made no attempt to wash loose activity from the floors. Air samples were taken with lapel samplers and a series of fixed samplers, located either near the floor (at 15 cm above the floor surface) or far above the floor at heights reaching 175 cm above the floor surface. Walking on the surface was done at 14 steps/minute, 36 steps/minute, and 100 steps/minute while blowing air with a hair dryer directed at the floor. Jones, 1967, results are summarized in Table 2 below.

Table 2 - Results of Jones, 1967, Study

Condition	Min RF, 10^{-6} m^{-1}	Max RF, 10^{-6} m^{-1}	Median RF 10^{-6} m^{-1}
Pu Oxide, 14 steps per minute	0.6	20	1.27
Pu Oxide, 36 steps per minute	1	177	16.2
Pu Nitrate, 14 steps per minute	0.3	1.33	0.64
Pu Nitrate, 36 steps per minute	1	16.2	3.02

4.5.2 Evaluation

The fixed air sampler results were reported as the average for 15 individual samples taken at heights from 15 to 175 cm above the floor. Using values that were determined near the floor where airborne concentration is higher than the breathing zone will tend to overestimate the RF for decommissioned sites. Personal air sampler results, where available, averaged 36 percent of the room air samples. The results of this experiment are not sufficiently representative of the building occupancy scenario for a decommissioned facility because they were done with freshly deposited solution and loose particles on smooth surfaces. This will cause the measured RF to be overestimated for the building occupancy scenario, which assume cleaned surfaces.

4.6   Nardi, 1999, Data

4.6.1 Description

Since the issue of dose estimates for the contamination in a building-occupancy scenario and the related issue of data for resuspension factor estimates had been raised at a series of public workshops, the NRC staff requested contributions of additional data on RF. In response, A. Joseph Nardi, a supervisory engineer with Westinghouse Electric Company, presented significant resuspension data at the NRC's public Workshop on Decommissioning, held on March 18-19, 1999. Mr. Nardi also provided the NRC on October 28, 1999, (Letter from A.J. Nardi, Westinghouse Electric Company, to N. Eisenberg, then NRC staff, now retired) with supplemental information on the data presented at NRC's workshop.

In Nardi's study, measurements of total surface activity were compared with airborne activity at a "Pump Repair" facility undergoing decommissioning. The facility consists of the main building, which is an open high-bay area 49. 6 m long x 12.2 m wide x 9.1 m high (142.5 ft. long x 40.0 ft. wide x 30 ft. high) and a tank room 14.6 m long x 3.7 m wide x 5.5 m high (48 ft. long x 12 ft. wide x 18 ft. high).

1 There was no forced air circulation within the building, and the only ventilation came through
2 open doors and convection from space heaters. HEPA filters were used locally on the equipment
3 during the shot-blasting operation of the floors when dust levels from rigorous cleaning activities
4 were locally high. The filters were placed locally on the equipment by the manufacturers
5 because of OSHA requirements for the protection of personnel. They were never used as part of
6 the facility ventilation system and no local HEPA filters were used during other decommissioning
7 operations (e.g., other than shot-blasting). Furthermore, the filters placed on the shot-blasting
8 equipment were characterized by very low air-flow rate and intended to reduce scattering of
9 particles from the floor caused by the shot-blasting process as required by OSHA. The impact of
10 the filters on the overall RF within the facility is minimal because it is localized and the air-flow
11 through the filter is rather small compared with the air-flow of the facility.

12 The radionuclides of primary interest for this facility included Co-60 and Cs-137. Air sampling
13 was conducted using 13 fixed-head air sample stations. The air sampling change frequency was
14 1-7 days depending on operational considerations. A typical flow rate of air samplers was
15 approximately 17000 cm^3/minute (0.6 ft^3/minute).

16 The data included 377 air samples, representing two data sets. A total of 247 samples were
17 collected for the first data set and 130 samples were collected for the second data set. The first
18 data set was generated using measurements taken before and during the initial decontamination
19 activities. Although there were no plant operations being performed in the period prior to
20 decommissioning, there was sufficient human activity at the site in the vicinity of the air samplers
21 to warrant inclusion of the data collected. Three different activities were performed while taking
22 these measurements during the decommissioning period; the removal of equipment from the
23 room, a one-pass shot-blasting of the floor, and waste packaging. The first data set samples
24 were collected from 13 different stations within the facility. The average air concentration of the
25 247 data points was 4.66 x10^{-8} Bq/ml (1.26 x 10^{-12} µCi/ml). The total surface activity
26 measurements under similar conditions were reported to be 26.7 Bq/100 cm^2 (1.6 x 10^5 dpm/100
27 cm^2). Thus, the nominal RF value before and during decommissioning activities is 1.7 x 10^{-7} m^{-1}.
28 The data also included surface contamination measurements from 29 locations, both before and
29 after floor contamination.

30 The second data set was generated using measurements taken after decommissioning while the
31 facility was essentially in a shutdown mode with minimal physical activities taking place. The
32 samplers for the second data set were located at the same 13 stations. The average RF value
33 corresponding to these condition are 4.2 x 10^{-8} m^{-1}.

34 **4.6.2 Evaluation**

35 The first data set represents typical facilities that are undergoing decontamination. However, the
36 conditions of driving forces causing resuspension were more aggressive than those conditions
37 representing a typical light-industry scenario. In addition, ventilation was minimal. Therefore,
38 depletion of the source-term were ineffective leading to more airborne concentrations and
39 consequently RF values, for the measured facility, higher than would be anticipated for a
40 decommissioned facility. On the other hand, the second data set represented less aggressive

driving conditions for resuspension than expected for the light industry scenario. However, ventilation was nearly static which causes a lesser depletion of the source-term and subsequently increase in resuspension. Therefore, the data for the second set may lead to an underestimate of the RF corresponding to the building occupancy scenario, and were therefore not used. The average RF derived from data taken during the post-decommissioning phase may underestimate the mean value for a light industrial scenario.

4.7 Ruhter and Zurliene, 1988, Data

4.7.1 Description

This study presented a brief discussion of airborne concentrations relative to surface contamination in the Three Mile Island, Unit 2 (TMI-2) auxiliary building during cleanup activities about 6 months after the accident. The principal source of airborne particulate radioactivity was resuspension of radioactive contamination which had been deposited on the surfaces. The report did not provide much data that can be broken down into subsets of measurements representing different facilities or various occupancy conditions. A maximum particulate concentration of 220 Bq/m^3 (5.94 x 10^3 pCi/m^3) was reported. Contamination levels on the floors were reported as high as 2000 - 4000 Mbq/m^2 (54 - 108 mCi/m^2). These values correspond to RF values in the range of 0.055 x 10^{-6} to 0.11 x 10^{-6} m^{-1}. However, both the surface and airborne values reported were maximums, so the resulting RF could be in error. The authors stated that "...a resuspension factor on the order of 10^{-8} cm^{-1} (i.e., 1x10^{-6} m^{-1}) would be expected from undisturbed surfaces, and would result in airborne concentrations similar to those observed...", but provided no additional information to support their affirmation.

4.7.2 Evaluation

Building surfaces had not been cleaned; thus, the test conditions could lead to an estimate of the RF higher than expected for decommissioned facilities. There are no specific measurements available for breaking the above data range into individual measurements representing different conditions.

4.8 Spangler, 1998, Data

4.8.1 Description

David Spangler, of the BWX Technologies, Navy Nuclear Fuel Division, presented resuspension data at the NRC's public Workshop on Decommissioning, held on December 1, 1998. These data were later amended in a written communication (Olsen, 2000). The RF was measured in a uranium storage area, the central storage vault, during handling of containers of uranium at an operating uranium fuel fabrication plant. Surface residual radioactivity concentrations were measured for both floors and uranium containers, both of which could contribute airborne activity from resuspension. Fixed air samplers collected approximately 1000 airborne radioactivity samples for the storage area of the fabrication plant. Approximately 4000 wipe test samples were also collected for the same facility. The data were generated over 12 months, during 1995. It appears that the facility meets the definition of a light-industry scenario. The three-year average RF values were: 4.25 x 10^{-7} m^{-1}, 7.79 x 10^{-6} and 8.97 x 10^{-7} for

1	fixed-air measurements, breathing-zone (BZ)measurements for averages < 6 hours, and BZ
2	measurements for averages ≥ 6 hours, respectively.

4.8.2 Evaluation

4	These data could represent a decommissioned facility, in terms of the expected driving forces of
5	a light-industry scenario. However, the airborne concentration may be exaggerated, because of
6	the resuspension from contaminated surfaces of containers and movement of such containers.
7	This is especially true with the BZ measurements, which tend to overemphasize the intake of
8	particles that were created by the mechanical operations such as opening or moving containers.
9	The third value reported above is for measurements with at least a 6 hour averaging time, and
10	are much lower than the peak BZ values of RF. The data also show that fixed contamination
11	varies over a relatively a small range $3.4 \pm 2.7 \times 10^2$ Bq/100cm^2 ($2.04 \pm 1.6 \times 10^4$ dpm/100cm^2)
12	whereas airborne concentration varies by approximately a factor of 6. As with the other data
13	used in the present study, the 12 monthly values reported may be combined into a single annual
14	average RF for this site.

15	There was surface activity, on the containers being moved, that would not be present in the
16	building occupancy scenario. This could cause the RF from this study to overestimate the RF at
17	decommissioned sites. Therefore, we will include only the RF values based on fixed air
18	samplers, and ignore the BZ data. Overall, the data appear to be representative of the building-
19	occupancy scenario and can be used for estimating the RF.

4.9   Summary of Data Used for Revising the RF

21	Although we have performed an extensive literature search, the number of measurements of
22	indoor RF is limited. Furthermore, the few measurements that are reasonably representative of
23	the building-occupancy scenario contain factors that will likely lead to an overestimate of RF.
24	There is currently not enough information to estimate the magnitude of this likely over-estimation.
25	Therefore, we must use our judgment to develop a distribution that we believe appropriately
26	reflects conditions for the building-occupancy scenario.

27	Table 3 shows ranges of RF values reported for two main types of particles or surface
28	contaminants. As can be seen in Table 3, the RF is significantly dependent on whether or not the
29	particles were freshly deposited on the surface. The studies involving freshly deposited
30	contamination have a high fraction of loosely bound particles; whereas the studies involving
31	operating facilities or those undergoing decommissioning have a significantly lower fraction of
32	loosely bound particles. None of the sites in the first category represent decommissioned
33	facilities in the respect that the surfaces had been decontaminated[2]. We anticipate that most
34	owners of facilities undergoing decommissioning will wash or otherwise clean contaminated
35	surfaces to comply with ALARA requirements of 20 CFR 20.1402. The approach used in
36	NUREG/CR Volume 3 (Beyeler, et al., 1999) was based on data from Fish, et al., (1967) and
37	Jones and Pond (1967), involving freshly deposited material. This approach assumed that the
38	RF would be proportional to the "loose" fraction as measured by a wipe measurement. This
39	proportionality was assumed to hold even if the fraction of "loose" particles was as low as a
40	couple of tenths of a percent, as would be typical at a decommissioned facility that had been

---

[2]The post-decommissioning Nardi data would qualify, but were not included in the final
assessment of RF.

16

washed. Rather than basing the RF on a study using freshly deposited material and proportionally reducing the RF by an assumed factor accounting for the fraction of loose particles likely to be present at decommissioned facilities, as was done previously (Beyeler, et al., 1999), the approach in this paper is to use data more directly applicable to decommissioned facilities.

Three sets of data (Breslin, 1966; Nardi, 1999; and Spangler, 1998) appear to be most applicable to estimating RF for decommissioned facilities. The measurements of Eisenbud, 1954, and Ruhter, 1988, appear to be less applicable, but still usable. Data from these five studies were used in this paper to develop an alternate distribution for RF.

5.0 DEVELOPMENT OF AN ALTERNATE ESTIMATE FOR RF

This section describes the statistical methods used to: (1) analyze the data described in Section 4; (2) develop an alternate RF PDF; and (3) select an appropriate default value for RF, for certain circumstances.

The approach used was a statistical analysis of all available data to evaluate the two empirical distributions (normal and log-normal) of the mean value of RF for each facility considered applicable. From the distribution, we report the 90th percentile value of the RF. Because of the sparsity of data, we also considered (but ultimately did not use) a tolerance limit to calculate the 90th percentile PDF value, with a 95th percentile confidence. In addition, an analysis of the time-dependance of the airborne concentration was performed for the Breslin and Nardi data sets to correct the RF values for worker occupancy times.

Table 3: Summary of RF Data Applicability

Study	Range of Resuspension Factor Values (m^{-1})
Freshly Deposited Contamination	
Fish, 1967	9.4 to 710 x 10^{-6}
Ikezawa, 1980	2.3 to 180 x 10^{-6}
Jones, 1967	0.3 to 177 x 10^{-6}
Cleaned or Aged Contamination	
Breslin, 1966	0.33 to 2.08 x 10^{-6}
Eisenbud, 1954	0.1 to 0.5 x 10^{-6}
Nardi, 1999	0.067 to 0.227 x 10^{-6}
Ruhter, 1988	0.055 to 0.11 x 10^{-6}
Spangler, 1999	0.425 x 10^{-6}

5.1   Correction Factor for Time Variation of Airborne Concentration

One consideration in the use of available data on airborne concentrations at indoor facilities is that the filters used to collect these data are generally in operation all the time, but workers are exposed only during the time they are there. These data need to be corrected to estimate RF because invariably the airborne dust load would be smaller when there was no activity within the buildings. The worst case would be that the airborne dust load falls to zero concentration after the workers leave for the day. In this case, the RF should be adjusted upward by a factor of 4.2, for a 40-hour work week; i.e., the ratio of 168 hours to 40 hours. However, the dust levels do not fall to zero after workers leave because the finest particles settle slowly, and there are other factors such as ventilation and natural convection that lead to a continual suspension of part of the dust.

Consider that the facility can be represented by a well-mixed room of volume V m^3. During worker activities, dust is generated in the room at a rate $W(t)$ grams/hour. Dust is removed from the room at a rate λCV grams/hour where λ is a first-order removal rate describing all removal mechanisms, including purging by ventilation and settling. The concentration C of dust in the room can be described by the first order ordinary differential equation:

$$\frac{dC}{dt} = \frac{W(t)}{V} - C\lambda \qquad\qquad (2)$$

This equation can be solved to calculate the concentration, and therefore the exposure rate in the room. The correction factor for worker duty cycle, DS, can then be calculated as the ratio of the average concentration during the time that the workers are present to the average concentration for the entire 168 hour week.

The Breslin (1966) data show the concentration of radioactivity versus time for nine samplers. Figure 2 shows the calculated RF values at 9 stations within the plant at four separate times. These times represent different periods around operational activities and show how airborne concentrations increase by these activities and decrease when they stop. The lines connecting the time points should be considered to be visual aids only, rather than an indication of the behavior between measurement times.

Analysis of the Breslin data indicate rather clearly that the airborne concentrations persist for considerable periods of time, and that the higher concentrations change at a faster rate than the lower concentrations. The most likely explanation for this observation is that the higher concentrations represent larger-sized particles, that must have been generated or suspended by more energetic processes than the finer particles. This observation is relevant to the choice of the RF value to be used for three reasons: (1) particles in the small-sized category are more likely to be the type generated in a light-industrial scenario, (2) small particles are more respirable, and (3) small particles are more likely to persist between work shifts in the building. The observed persistence of airborne concentration diminishes, but does not eliminate, the importance of daily activities in elevating the airborne concentration.

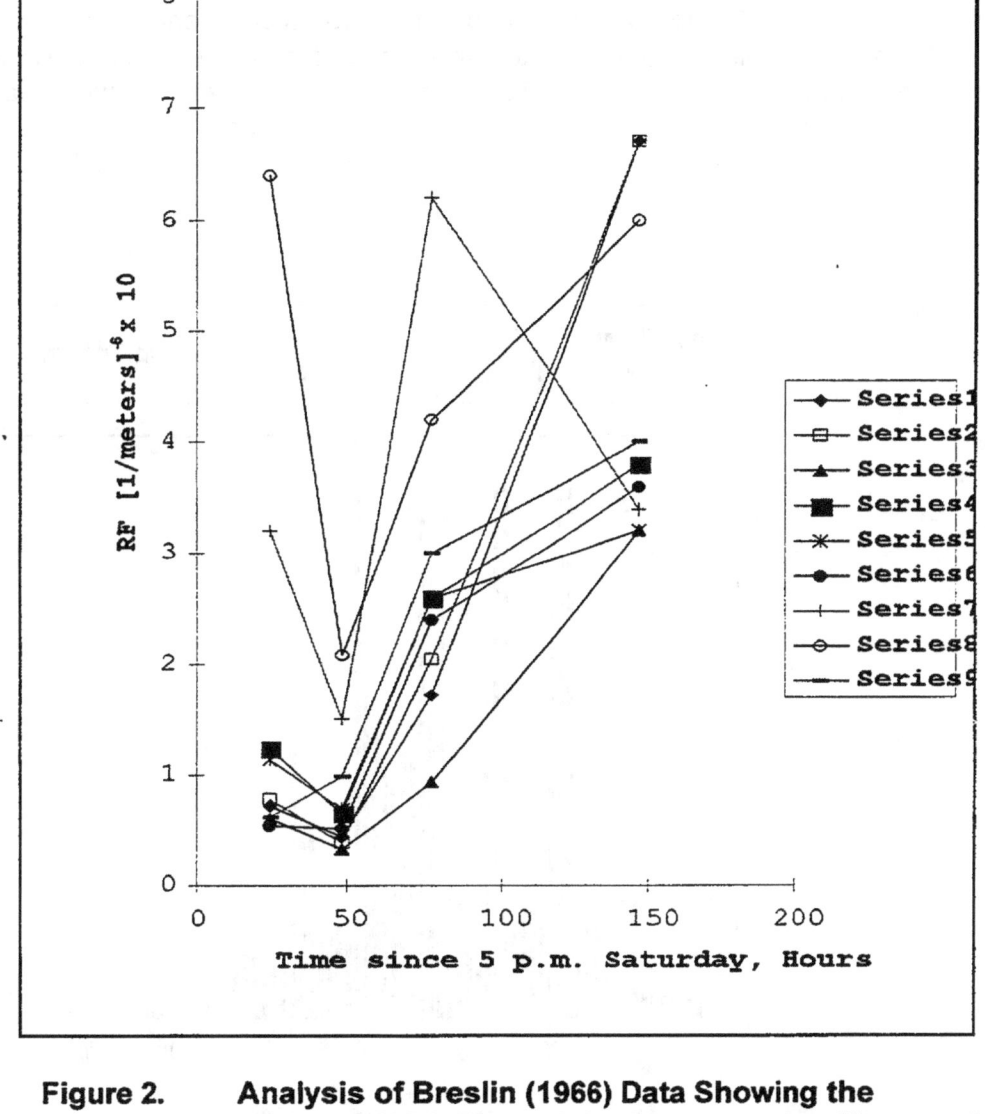

Figure 2.   **Analysis of Breslin (1966) Data Showing the Variation With Time of the Concentration Measured at Different Sampling Locations.**

1   An average decay rate from the first part of the Breslin data is λ = 0.029 hr^{-1}. If the average value
2   applied, concentration would fall roughly to half the highest value by the start of the next morning
3   (assuming a single 8-hour work shift). Excluding the two highest measurement stations leads to
4   a much smaller removal rate of 0.00946/hour. An alternative estimate of
5   λ = 0.022 to 0.054 hr^{-1} with an average value of λ = 0.0378 hr^{-1} can be based on the assumption
6   that the airborne concentrations in the Breslin facility are cyclical, but do not vary from week to
7   week.

19

1 Figure 3 shows air sampling results for the Nardi site. These data are less descriptive than the
2 Breslin data, and it was not possible to estimate the decay rate *a priori*. Instead the model
3 represented by Equation 2 was run for different values of λ, and the concentrations as measured
4 by the air filters calculated under the assumption that filters were changed at
5 8:00 a.m. on Monday and 5:00 p.m. on Thursday. There were, therefore, two assumed periods
6 of averaging; 1) the 82 hours during which there was worker and decommissioning activity, and
7 2) the 86 hours of minimal activity after workers left for the week. A qualitative comparison of the
8 measured (Figure 3) and predicted (Figure 4) average concentrations indicated that a value of λ
9 = 0.05 hr^{-1} was most appropriate for the conditions at this site.

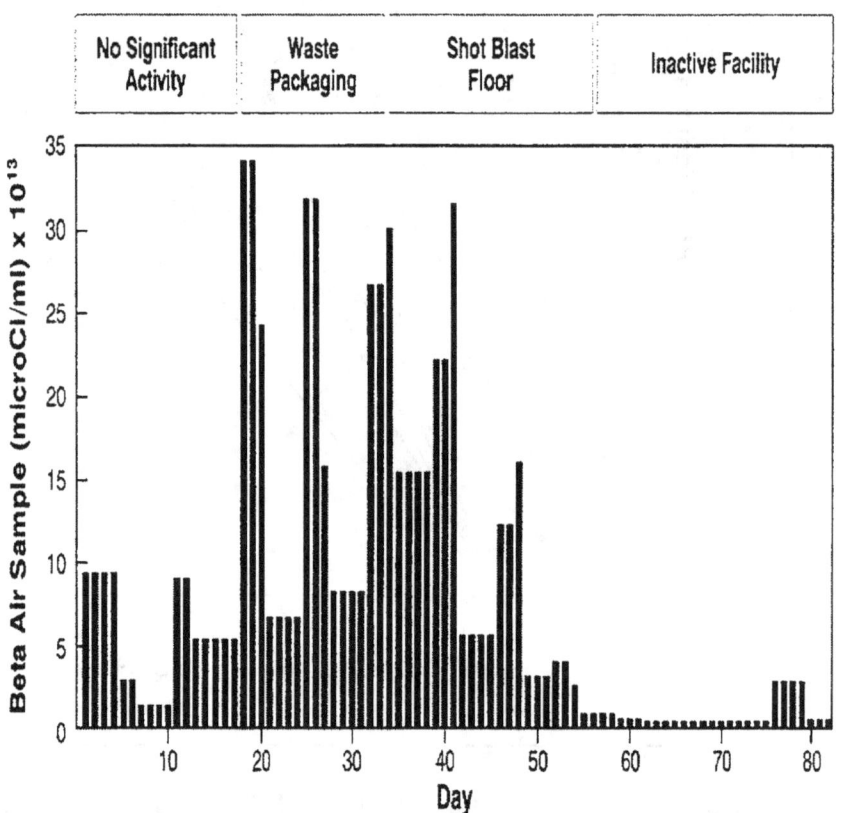

**Figure 3.   Beta Air Activity Sample Results for
Westinghouse Active/Inactive
Decommissioning Facilities**

20

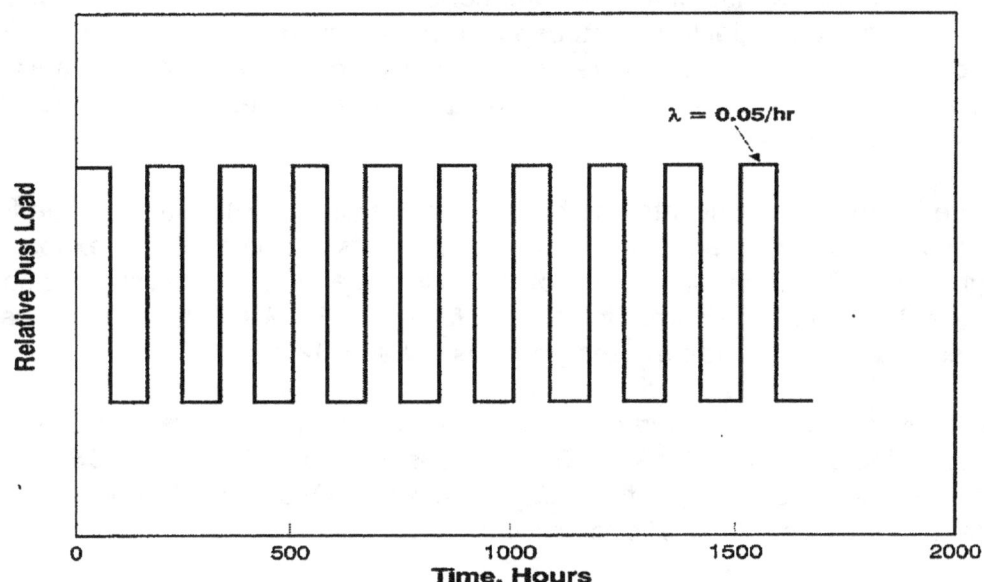

Figure 4.   Predicted Dust Load for Nardi, (1999) Data

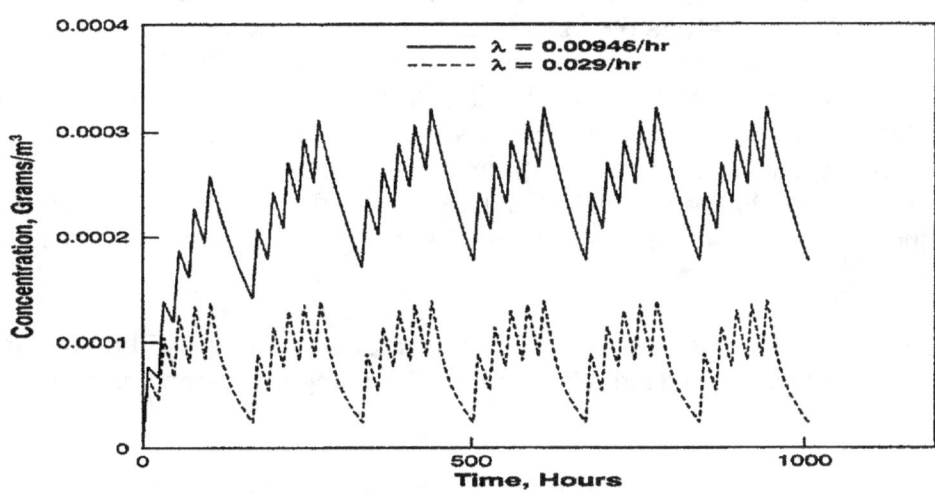

**Figure 5.   Predicted Variation of Dust Concentration (gm/cm^3)
Using Breslin Data (1966)**

We estimated the correction factor for worker occupancy by numerically integrating Equation 2, and calculating the ratio of the average concentration during worker exposure to the weekly average, after the initial transient response for the system has died away. For the Breslin site, the dust source term, $W(t)$ was represented as a square wave input that was 1.0 gram/hour for 8 hours a day, and 0 grams/hour for the next 16 hours, repeated for 5 days, followed by an input of 0 grams/hour for the 48 hour weekend.

Figure 5 shows the concentration buildup for Breslin (1966) data with time for a 6 week cycle starting with zero concentration in the air. Workers are present and exposed only on the upward segments of the "sawtooth". The volume is immaterial for calculating the correction factor DS. For $\lambda = 0.0378$ hr^{-1}, the correction factor $DS = 1.2$. For $\lambda = 0.00946$/hour, perhaps more representative of respirable-sized particles, $DS = 1.02$.

For the Nardi's data, we made similar assumptions, but the workers were exposed for 4 ten-hour days. Using a decay factor $\lambda = 0.05$ hr^{-1} leads to a correction factor DS = 1.5. Interestingly, the correction factor is larger for the 4-day work week. Assuming 8 hour days, 5 days a week led to a smaller correction factor DS = 1.28 for this site.

5.2   Statistical Analyses of the RF

5.2.1   Adjustments to Data

Data available for statistical analysis consist of the average RF values for the five sites (Nardi, Breslin, Spangler, Ruhter and Eisenbud). Each of the average RF values was adjusted upward by an "occupancy" factor. Occupancy correction factors were available only for the Breslin and Nardi data. The average RF values for the Breslin and Nardi sites were adjusted by a factor of DS = 1.2 and 1.5, respectively. The average RF values for the for the remaining three sites were adjusted by a factor of 1.35, which is the average for the Breslin and Nardi corrections. We feel that these correction factors are conservative, mainly because the filters that were used to collect the airborne particles probably captured a significant fraction of larger particles, which settle faster, and lead to the calculation of higher λ, and thus higher DS. This might be especially true for the Nardi data, which included periods of high-energy operations such as shot-blasting of surfaces.

The corrected RF data for the five sites is shown in Table 4. The cumulative probability for normal and lognormal distributions of the RF using the mean values of five facilities is shown in Figure 6.

Table 4:   Mean Values of RF for Each Site

Site Reference	Mean RF, 10^{-7} m^{-1}	Mean RF, 10^{-7} m^{-1}, Adjusted for Occupancy
Nardi (1999, Decommissioning)	1.71	2.565
Spangler (2000)	4.25	5.734
Ruhter and Zurliene (1988)	0.825	1.114
Breslin (1966)	8.44	10.13
Eisenbud (1954)	3.07	4.145

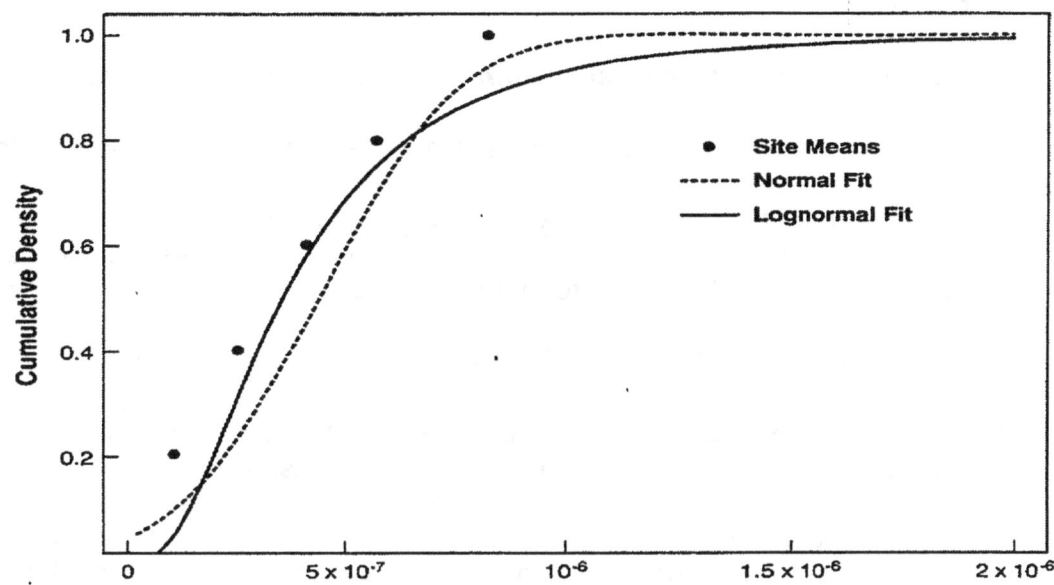

Figure 6.   **Cumulative Probability Distribution (Normal &Lognormal) of the RF Using Mean Values of Five Facilities**

5.2.1.1 Tolerance Limits

The statistical confidence in the estimated value of the 90th percentile RF can be calculated for the size of the sample under the assumption that RF (or its logarithm) is normally distributed. The confidence in the value of RF can be stated: "At least 90 percent of the values of RF would be less than μ - k s with a confidence of 95 percent", where μ is the sample mean of RF and s is the sample standard deviation of RF. A similar statement would apply to the logarithm of RF.

Tolerance is an issue because we are using a small amount of data to estimate the PDF describing the variability of RF over the various NRC decommissioning sites. The variability among various sites is an aleatory uncertainty, while the tolerance describes how certain we are of the knowledge base, i.e. an epistemic uncertainty. If we had a large number of data, say hundreds to thousands of samples, to estimate the PDF, the tolerance bands around the nominal value would be small. However, with sparse data, the tolerance bands can be significant. The methodology for obtaining the 90th percentile of the dose distribution assumes that the PDF's are precise, i.e., (1) estimation error is not explicitly represented; and (2) the derived PDF's do not appear to take into account how much data are available (almost always sparse data) to make the estimate. Since the rest of the methodology for obtaining screening values does not consider the amount of data available to estimate the input variable PDF's, it would be inconsistent to take

23

this into account for resuspension factor. Furthermore, because of the nature of the data used, which we believe overestimates the value of RF, and because features of the model (e.g., no depletion by ventilation) also tend to overestimate dose, use of the nominal value is deemed appropriate. However, consideration of estimation error in dose modeling, (perhaps in risk analysis in general) may be a topic that needs further study within the entire context of regulatory decision-making.

5.2.1.2 Consideration of the Post-decommissioning Data from Nardi

The average value of RF calculated from the post-decommissioning Nardi data are about 1/4 those calculated from the decommissioning results. This result could be used to lower the estimates presented from the other data by a similar factor. However, the post-decommissioning data may be unrepresentative of a light-industrial scenario. Consequently, we decided that this factor will not be used in the estimate of the RF distribution.

5.2.2   Statistical Analysis of Site Mean Results

Although some of the data indicate that there are likely to be significant variations in airborne concentrations from place to place, one may wish to consider that there is an overall effective RF for each site. Occupants of the buildings are likely to move around; therefore they are exposed to a variety of potential resuspension conditions rather than one. Under these assumptions, it is appropriate to use the site average of the RF's as a sample representing the variability of RF across the population of NRC licensees.

6.0   SUMMARY, CONCLUSIONS, AND RECOMMENDATIONS

6.1   Summary

The authors of this report followed the following approach to reevaluate the PDF and nominal value of the indoor resuspension factor for use in screening evaluations for the license termination rule:

1.   Modeling the building occupancy scenario for contamination with α-emitters resulted in doses higher than those obtained with other standard codes; in some cases the indicated cleanup levels were below detectable limits.

2.   Evaluation of the models used in the building occupancy scenario indicated that the resuspension factor parameter, both the PDF and default value, was the primary cause of this result.

3.   Examination of the basis for the PDF used previously indicated the data were obtained under conditions that did not match very well the conditions anticipated at decommissioned facilities.

4.   The technical literature was reviewed to obtain further data for indoor resuspension factors.

24

1 2 3 4	5.	Participants at NRC's public workshops on implementation of the License Termination Rule were asked to provide additional data on indoor resuspension factors. Additional data were provided by D. Spangler of BWX Technologies, and A. Nardi, of Westinghouse Power Corporation.
5 6 7	6. 7.	A total of eight sets of data were evaluated for applicability to decommissioned facilities. Five sets of data were deemed applicable enough to use to quantify the PDF for the indoor resuspension factor.
8	8.	Data were corrected to account for the fraction of the time workers occupy the site.
9	9.	We performed several statistical analyses:
10 11	a.	A determination of the PDF for mean values of RF from each of the five studies (5 separate estimates of RF).
12	b.	Evaluation of different functional forms of the PDF (lognormal and normal).
13	c.	Determination of the nominal value of the 90th percentile of the PDF.
14 15	10.	These analyses were evaluated and a preferred choice of PDF and the 90th percentile of that PDF were chosen.

6.2 Conclusions

17   The additional information, both in the literature and provided by two facilities, appears to be an
18   improved basis to estimate indoor resuspension factor. Nevertheless, these data have certain
19   limitations, most of which relate to the applicability to decommissioned facilities of the conditions
20   under which data were obtained. These limitations include:

21 22 23	1.	Interference from Operations. An apparent elevation of air concentrations occurred in some cases (Breslin, Eisenbud, and Nardi) when measurements were made in facilities where operational activities introduced radioactive material directly into the air.
24 25 26 27 28 29	2.	Different Resuspension Forces. In some cases, the resuspension forces were simulated (Breslin); in other cases, the resuspension forces were absent, because the facility was not in use at the time of measurement (Nardi). In the former case, the measured resuspension factor could be higher or lower than that in a decommissioned facility, depending on the nature of simulated activity; in the latter case, the measured resuspension factor could be lower than that in a decommissioned facility.
30 31 32 33 34 35	3.	Location of Measuring Instruments. There are several factors with the location and type of measuring instruments. Data from fixed air samplers were preferred because they better indicated levels of respirable dust than breathing zone, lapel samplers. Location is also important. Lapel samplers are at the correct height, but tended to reflect contamination levels from equipment operation rather than resuspension. Samples taken close to the floor were considered inappropriate for data on respiration.

4. Condition of the Contamination. In a decommissioned facility, it is anticipated that the contaminated surfaces will have been cleaned, so loose particulate matter harboring contamination will have been removed. However, as surfaces are subject to wear and other forces, some of this "fixed" contamination may become loosened. Alternatively, maintenance activities such as waxing floors or painting surfaces, may more firmly fix residual contamination. Some tests (Jones and Pond, 1967, Fish et al, 1967, Ikezawa, et al., 1980) used freshly deposited material, which probably overestimates RF.

5. Other Conditions of Measurement. Other conditions existing during the time that measurements were made may also influence the degree to which the data obtained apply to a decommissioned facility. Ventilation at the contaminated sites was not well-characterized, and it is difficult to determine how well ventilation expected in decommissioned facilities corresponds to the data. Another possible example is the use of HEPA filters during decommissioning operations. Nardi (1999) reports that such filters were in use during some of the decommissioning operations, but only to protect workers near the operating machinery. We decided that the use of the filters in this case did not generally decrease the airborne dust load since they acted only on the operating equipment producing the dust, and not on resuspended dust.

In summary, the available data are not perfect, but they do provide the best insight available at the present time into an estimate of the PDF for resuspension factor. Overall, the authors of this report believe these data provide an overestimate of the distribution of RF's likely to exist at decommissioned facilities.

The methodology used to develop default parameters for the DandD code presumed that the PDF's describing the variability of parameters among NRC-licensed facilities was precise, but sparse. Even though this uncertainty may be significant, we conclude that the "best estimate" of the PDF should be used for screening analyses. Two important reasons for choosing this strategy are: (1) the exposure scenario, dose models, PDF's for other parameters, and the data supporting quantification of the PDF for RF are all believed to contain significant conservatisms, which argue against using the extra measure of conservatism introduced by insisting on high confidence in the results (i.e., using tolerance); and (2) the remainder of the DandD screening analysis uses PDF's that do not consider estimation uncertainty. Therefore, consideration of tolerance for RF only would be inconsistent and would introduce more unnecessary conservatism for radionuclides affected by resuspension.

We deemed it appropriate to base the PDF for RF on the 5 data points representing the site means, adjusted for worker occupancy, because: (1) workers may move around a facility and be exposed to a variety of air concentrations; and (2) the regulation is written to protect the average member of the critical group. We fitted the five site data to a normal and a lognormal distribution. Since there were only five data points, we felt that it was appropriate to use the "maximum likelihood" approach (Benjamin and Cornell, 1970) to estimate the distribution rather than a statistical (i.e., "unbiased") approach. The difference between the two approaches is that the estimated standard deviation in the maximum likelihood approach is smaller by the ratio $\sqrt{(N-1)/N}$. This smaller standard deviation will lead to a slightly smaller value for the 90th percentile of the distribution, which is used as the suggested regulatory criterion for RF. Parameters of the normal and lognormal distributions are given in Table 5 for the maximum likelihood fits. Figure 6 shows the two distributions. Also shown in this figure are the original data plotted as an empirical distribution, with the smallest value equal to the 10th percentile and the largest as the 90th percentile. Although both the normal and lognormal distributions are reasonable fits to the data, the normal distribution has the disadvantage of allowing negative values of RF, which is not physically possible. In addition, the lognormal fit is the more conservative choice at the 90th percentile RF.

Table 5: Parameters for Normal and Lognormal "Maximum Likelihood" Models of RF Data

Statistical Model	Sample Mean	Sample Standard Deviation	90th Percentile RF
Normal Fit to 5 site mean RF's	4.74×10^{-7} m^{-1}	3.11×10^{-7} m^{-1}	8.7×10^{-7} m^{-1}
Lognormal Fit to 5 site mean RF's	$\log_{10} = -6.433$	$\log_{10} = 0.3247$	9.6×10^{-7} m^{-1}

6.3   Recommendations

We make the following recommendations:

1. The PDF given in Section 6.2 should be implemented in the DandD code. For the building occupancy scenario with the additional condition that the dose is dominated by inhalation of a single radionuclide, the nominal 90th percentile of the lognormal fit for RF , (i.e. 9.6×10^{-7} m^{-1}), may be used. For situations where other pathways (e.g., direct exposure) are significant, this PDF must be processed through the DandD code screening methodology.

2. Because of the paucity of data and the incompatibility of the conditions under which it was obtained and conditions anticipated for decommissioned facilities, consideration should be given to conducting research to obtain more data directly applicable and representative of facilities whose licenses are to be terminated by NRC.

3. Sparse data for the estimation of the properties of a distribution can lead to significant uncertainties in the properties of the distribution (e.g., the mean, the standard deviation, the 90th percentile). Consideration is usually not given to this type of uncertainty in the PDF's used for dose estimates, performance assessments, and probabilistic risk analyses. The NRC staff should investigate the impact of estimation uncertainty and how it may affect regulatory decisions in a risk-informed, performance-based regulatory context.

7.0 REFERENCES

Abelquist, E. W, W. S. Brown, G. E. Powers, A. M. Huffert, "Minimum Detectable Concentrations with Typical Radiation Survey Instruments for Various Contaminants and Field Conditions," NUREG-1507, 1998.

ASHRAE 89"Ventilation for Acceptable Indoor Air Quality," ANSI/ASHRAE 62-1989, American Society of Heating, Refrigerating and Air-Conditioning Engineers, Inc., Atlanta, 1989.

Benjamin, J.R., and C. A. Cornell, *Probability, Statistics, and Decision for Civil Engineers*, McGraw-Hill, 1970

Beyeler, W.E., W.A. Hareland, F.A. Durán, T.J. Brown, E. Kalinina, D.P. Gallegos, and P.A. Davis, *Residual Radioactive Contamination from Decommissioning, Parameter Analysis.* Draft report for comment, NUREG/CR-5512, Vol. 3. U.S. Nuclear Regulatory Commission, Washington, DC., 1999.

Beyeler, W. E. et al., "Review of Parameter Data for the NUREG/CR-5512 Building Occupancy Scenario and Probability Distributions for the DandD Parameter Analysis," Draft Letter Report dated January 30, 1998.

Breslin A. J., A. C. George, and P. C. LeClare, "The Contribution of Uranium Surface Contamination to Inhalation Exposures," HASL-175, 1966.

Brodsky, A., "Resuspension Factors and Probabilities of Intake of Material in Process (Or 'Is 10^{-6} a Magic Number in Health Physics?')," *Health Physics, 39*, 992.

Cember, H., 1996, *Introduction To Health Physics*, McGraw-Hill, New York, NY

Corn, M. and F. Stein, "Mechanisms of Dust Resuspension" in *Surface Contamination: Proceedings of a Symposium Held in Gatlinburg Tennessee, June 1964*, B. R. Fish, ed., page 45, Pergamon Press, Oxford, 1967.

EPA 1988, *Limiting Values of Radionuclide Intake and Air Concentration and Dose Conversion Factors for Inhalation, Submersion, and Ingestion*, Federal Guidance Report No. 11, Environmental Protection Agency, Washington, D.C.

Eisenbud, M., Blatz H., and E.V. Barry, "How Important is Surface Contamination?" *Nucleonics*, Vol. 12(8), 12, August, 1954.

Fish, B. R., R. L. Walker, G. W. Royster, Jr., and J. L. Thompson, "Resuspension of Settled Particulates," in *Surface Contamination: Proceedings of a Symposium Held in Gatlinburg, Tennessee, June 1964*, B. R. Fish, ed., page 75, Pergamon Press, Oxford, 1967.

Hinds, W.C., *Aerosol Technology - Properties, Behavior, and Measurement of Airborne Particles*, John Wiley & Sons, New York, 1982.

Ikezawa, Y., T. Okamoto, and A. Yabe, "Experiences in Monitoring Airborne Radioactive Contamination in JAERI," in Radiation Protection: A Systematic Approach to Safety - Proceedings of the 5th International Radiation Protection Society, p 495, Pergamon Press, 1980.

Jones, I.S. and S. F. Pond, "Some Experiments to Determine the Resuspension Factor of Plutonium from Various Surfaces," in *Surface Contamination: Proceedings of a Symposium Held in Gatlinburg Tennessee, June 1964*, B. R. Fish, Ed., page 83, Pergamon Press, Oxford, 1967.

Kennedy, W.E., "Provate Communication Between S. McGuire and W.E. Kennedy," 1999.
Kennedy W.E. and D. L. Strenge, NUREG/CR-5512, Volume 1, "Residual Radioactive Contamination from Decommissioning," 1992

Morton, H., "Interpretation of the Relation of Building Interior Surface to Airborne Radioactivity," Submitted at NRC Public Workshop on Decommissioning, January 1999.

Nardi, A.J., "Operational Measurements and Comments Regarding the Resuspension Factor, Presentation at NRC Decommissioning Workshop, March 18, 1999 and associated transcript.

Nuclear Regulatory Commission, "Final Rule on Radiological Criteria for License Termination," July 21, 1997, 62 FR 39058.

Ruhter, P.E.and W.G. Zurliene, "Radiological Conditions and Experiences in the TMI- Auxiliary Building," CONF-881-24-9, 1988.

Spangler, D.L., "Re-Suspension Factor Determination and Comparison Using Data from an Operating Licensed Facility," BWX Technologies, Inc., presentation at NRC Decommissioning Workshop, December 1, 1998 and associated transcript.

Walker, R.L. and B. R. Fish, "Adhesion of Radioactive Glass Particles to Solid Surfaces" in *Surface Contamination: Proceedings of a Symposium Held in Gatlinburg Tennessee, June 1964*, B. R. Fish, ed., page 61, Pergamon Press, Oxford, 1967.

8.0 APPENDIX A

Table A-1: SUMMARY OF RF VALUES BASED ON BRESLIN, et. al., (1966) DATA

Facility/Data Set		Calculated RF Values[1], (m^{-1})Multiplied By 10^6, Under Different Operational Conditions[2]			
		Condition "a"	Condition "b"	Condition "c"	Condition "d"
Assistant Press Operator Facility	Lapel Sampler of Worker 1	0.22	0.36	0.86	3.40
	Lapel Sampler of Worker 2	0.19	0.39	1.03	3.40
	Fixed Air Sampler	0.1	0.31	0.37	1.60
Rod Puller Facility	Lapel Sampler of Worker 1	0.33	0.62	1.30	1.90
	Lapel Sampler of Worker 2	0.35	0.57	1.30	1.60
	Fixed Air Sampler	0.26	0.27	1.20	1.80
Rod Straightener Facility	Lapel Sampler of Worker 1	0.75	1.60	3.10	1.70
	Lapel Sampler of Worker 2	1.04	3.2	2.05	3.00
	Fixed Air Sampler	0.49	0.31	1.50	2.00

---

[1] The RF is the ratio of airborne concentration of radioactive contaminant to the average surface activity. The airborne concentration was measured for each of the three facilities using two lapel samplers and one fixed air sampler. The surface activity values measured for each facility were given in Section 4.1.1. These values were multiplied by a factor of 2 because calibrations of alpha measurements of surface activity conducted in early studies (1954 - 1967) underestimated the total surface activity by a factor of two (Abelquist et. al., 1998).

[2] Condition "a" corresponds to measurements taken on Sunday with no operational impacts (e.g., airborne contamination introduced by operations had settled out of the air). Condition "b" corresponds to measurements taken on Saturday representing post-operation transient condition. Condition "c" represents initial operating transient conditions for measurements taken on Monday; and conditions "d" corresponds to measurements taken on Thursday with typical operation of the concerned facility (see Section 4.1.1 for details).

NRC FORM 335
(2-89)
NRCM 1102,
3201, 3202

U.S. NUCLEAR REGULATORY COMMISSION

BIBLIOGRAPHIC DATA SHEET

(See instructions on the reverse)

1. REPORT NUMBER
(Assigned by NRC, Add Vol., Supp., Rev., and Addendum Numbers, if any.)

NUREG-1720

2. TITLE AND SUBTITLE

Re-evaluation of the Indoor Resuspension Factor for the Screening Analysis of the Building Occupancy Scenario for NRC's License Termination Rule

Draft Report for Comment

3. DATE REPORT PUBLISHED

MONTH	YEAR
June	2002

4. FIN OR GRANT NUMBER

5. AUTHOR(S)

R.M. Abu-Eid, R.B. Codell, N.A. Eisenberg*, T.E. Harris, S. McGuire**

6. TYPE OF REPORT

Technical

7. PERIOD COVERED *(Inclusive Dates)*

8. PERFORMING ORGANIZATION - NAME AND ADDRESS *(If NRC, provide Division, Office or Region, U.S. Nuclear Regulatory Commission, and mailing address; if contractor, provide name and mailing address.)*

Division of Waste Management
Office of Nuclear Material Safety and Safeguards
U.S. Nuclear Regulatory Commission
Washington, DC 20555-0001

9. SPONSORING ORGANIZATION - NAME AND ADDRESS *(If NRC, type "Same as above"; if contractor, provide NRC Division, Office or Region, U.S. Nuclear Regulatory Commission, and mailing address.)*

Same as above

10. SUPPLEMENTARY NOTES

11. ABSTRACT *(200 words or less)*

The purpose of this study was to re-evaluate the resuspension factor (RF) parameter used in the screening analysis for demonstration of compliance, using the building occupancy scenario, with the radiological criteria in the license termination rule in 10 CFR 20, Subpart E. The RF is a highly sensitive parameter impacting the inhalation dose calculation. An RF parameter value of (1.42 E-4) /m was established for screening analysis (Beyeler et al, 1999). Assuming a 10% fraction of loose (removable) contamination, NRC staff selected a default RF value, for use in the inhalation dose calculation, of (1.42 E-5)/m. Based on this RF value, and using DandD code, the derived default concentration or surface activity screening limits for most radionuclides, particularly the alpha-emitters, were at background levels or far below the corresponding detection limits. In this study, NRC staff analyzed further literature data considering more realistic assumptions of the average member of the critical group in the building occupancy scenario and accounting for more recent actual RF field data collected for two decommissioning facilities. Based on the current analysis and re-evaluation, staff recommends using an RF value of (1.0 E-6)/m for use in the screening analysis of the inhalation dose calculation for the building occupancy scenario. Staff believes that the newly proposed RF default value is more realistic, than the current value in DandD code, and sufficiently conservative for conducting screening analysis.

12. KEY WORDS/DESCRIPTORS *(List words or phrases that will assist researchers in locating the report.)*

Resuspension Factor
Indoor Resuspension Factor
Decommissioning Screening Dose Analysis Default Parameters
Building Occupancy Scenario Parameters for Decommissioning

13. AVAILABILITY STATEMENT

unlimited

14. SECURITY CLASSIFICATION

(This Page)

unclassified

(This Report)

unclassified

15. NUMBER OF PAGES

16. PRICE

JUNE 2002

NUREG-1720
DRAFT

RE-EVALUATION OF THE INDOOR RESUSPENSION FACTOR FOR THE SCREENING
ANALYSIS OF THE BUILDING OCCUPANCY SCENARIO FOR NRC'S LICENSE
TERMINATION RULE

UNITED STATES
NUCLEAR REGULATORY COMMISSION
WASHINGTON, DC 20555-0001